PACKARD MOTOR CARS
1935 THROUGH 1942
PHOTO ARCHIVE

PACKARD MOTOR CARS 1935 THROUGH 1942

PHOTO ARCHIVE

Photographs from the Detroit Public Library's
National Automotive History Collection

Edited with introduction by Mark A. Patrick
Curator, National Automotive History Collection

Iconografix
Photo Archive Series

Iconografix
PO Box 609
Osceola, Wisconsin 54020 USA

Library of Congress Card Number 95-82098

ISBN 1-882256-44-1

96 97 98 99 00 5 4 3 2 1

Cover and book design by Lou Gordon, Osceola, Wisconsin

Printed in the United States of America

Book trade distribution by Voyageur Press, Inc. (800) 888-9653

PREFACE

The histories of machines and mechanical gadgets are contained in the books, journals, correspondence and personal papers stored in libraries and archives throughout the world. Written in tens of languages, covering thousands of subjects, the stories are recorded in millions of words.

Words are powerful. Yet, the impact of a single image, a photograph or an illustration, often relates more than dozens of pages of text. Fortunately, many of the libraries and archives that house the words also preserve the images.

In the *Photo Archive Series*, Iconografix reproduces photographs and illustrations selected from public and private collections. The images are chosen to tell a story—to capture the character of their subject. Reproduced as found, they are accompanied by the captions made available by the archive.

The Iconografix *Photo Archive Series* is dedicated to young and old alike, the enthusiast, the collector and anyone who, like us, is fascinated by "things" mechanical.

Packard Westchester, New York showroom, Ladies' Week 16-21 March 1936. Featuring (foreground) a 1936 Packard 120, Fourteenth Series, Model 120-B Two/Four-Passenger Convertible Coupe.

INTRODUCTION

Certainly any year would be a good point from which to launch a Photo Archive of Packard motor cars. Yet, certain events in 1935 would seem to make it a particularly good point in time. It was the middle of the Great Depression. President Roosevelt called for automotive manufacturers to coordinate their production efforts and release new models in the autumn of each year, as a way of fighting unemployment during the winter months. Before 1935, Packard introduced each new series according to engineering or styling advancements. From 1935, Packard models adhered closely to the model year designation—although Packard was noted for introducing their models in the late summer as opposed to autumn. That was almost certainly a signal to the customer that Packard was a step ahead of the rest of the industry.

It was also in 1935, that Packard introduced the Twelfth Series and put the Model 120 into production. Although the company had held its own through the early Thirties, if it was to survive a prolonged economic depression, it needed an automobile to compete toward the lower end of the market. Although the 120 was not an inexpensive automobile, it was priced to attract customers who sought the pride of Packard ownership. It did so—the company had customers by the bundles—nearly 80,000 Packard 120s were sold in two years.

Customers did not merely gain prestige. The Packard 120 was a very good automobile. For around a $1,000, the customer got a very respectable 282 cubic inch eight-cylinder in-line engine, independent front suspension, and hydraulic brakes. As this collection of photographs illustrates, perhaps the most striking aspect of the Packard 120 was just how well designed it was. Packard offered various body styles: business and sports coupes, convertible coupes, and a variety of sedans. In 1938, as the 120 moved upscale following the introduction of the six-cylinder 110, the company even added special Rollston bodies. Packard's advertising punctuated their arrival by proclaiming, "Real luxury in the lower priced field."

The economic depression dragged on. In 1937, the company reintroduced the six-cylinder Packard 110 in the Fifteenth Series. The 110 was a reflection of the proletarian trend in the industry to offer at least one model at a price that appealed to modest income buyers—the customer who might otherwise opt for a previously owned automobile. Whereas, the 120 was never inexpensive, the 110 certainly was—originally priced at $795. Packard kept subsequent price increases to a minimum. The 110 was priced in the affordable $800-900 range for several years.

Several writers have claimed that the six-cylinder Packard was not a great automobile, but none have claimed that it was a bad one. The company strove to offer quality and value, along with the Packard élan. Admittedly, the company cut corners. The engine was stripped of two cylinders, but it was not embarrassing and would rather easily push the 100 mph threshold. The wheelbase was shorter and the body styles were fewer, yet a convertible and sport coupe were available. The interior was not sumptuous, yet it retained a dignified, if modest, essence.

Still when one beholds a Packard 110, recognition is immediate and visceral. The Packard's proud vertical grille and the slightly paralleologram shaped hood louvers are styling features long associated with aristocratic elegance. Yet, the automobile was intended to be competitive in the mass market. Packard management was long recognized for their business acumen, and the Packard 110 was an outward representation of their keenness. The company adapted itself to the market, but the compromises it made were neither embarrassing nor ruinous of their reputation.

The Packard Senior Series Eights might be considered to have been the backbone of the company. While Packard did not sell as many Seniors as it did 120 and 110 models, one can claim the larger eight-cylinder motor cars were the Packards on Front Street—automobiles that reminded the public of the pinnacle Packard had achieved among motor cars. Senior Eights came in a variety of body styles that included custom LeBaron bodies. Engines were fairly straight forward—a 384 cubic inch, available in 1935 and 1936, and a 320 cubic inch available through the end of the Thirties. A 356 cubic inch engine was available from 1940 to 1942. As the photographs provide evidence, the Senior Eights were glorious automobiles.

The admiration and desire one feels toward the 12-cylinder Packard has a mythological-like quality about it. This is certainly due to more than the car's elusiveness—so few were built that the public rarely, and then only fleetingly, glimpsed one. The general public came to know the car through its advertising, a rather secondary but intense experience. The Packard Twelve was artistically rendered in non-primary—otherworldly—colors, and it was nearly always positioned in front of Greco-Roman inspired architecture. It was as if the beholder was called to associate the automobile with those distant ancient times—a modern chariot of the gods. Further, the Packard advertising slogan became the myth-mantra of the devoted—"Ask the man who owns one." The public experienced the Packard Twelve outside the pale of ordinary experience.

Even in our time, the Packard Twelve is seen by the public artificially—at the concours or in the automobile museum. The Packard Twelve is placed within a shrine. Devotions by the public are offered in the admission fees and gift shops. The judges, writers, and curators function as the automobile priesthood. This is not a negative paradigm, and should not be taken so. Rather, it is recognition that the automobile is much more than a mechanical device that conveys the user. In the instance of the Packard 12-cylinder, the emotional experience is particularly intensified. Would "graceful" be an appropriate description of the Packard? The images included here provide a concrete answer to the rhetorical.

The photographs appearing in *Packard Motor Cars 1935 through 1942 Photo Archive* are from the Detroit Public Library's National Automotive History Collection. We are nearly 50 years old. The mission of the NAHC is to retain and preserve the historical record of the automobile and other forms of wheeled transportation. Toward this aim, the NAHC has become the premier collection of its type. Our files include 350,000 pieces of sales literature and 250,000 photographs. The collection also houses biographical files, books, magazines, art, blueprints, owner's manuals, and personal papers of automotive pioneers and trailblazers. Much of this material is unique. Most importantly, it is in the public library and, therefore, is accessible to the enthusiast.

9

1935 Twelfth Series

Model 120 Five-Passenger Sedan. Photograph taken at the Packard Proving Grounds, Utica, Michigan.

Model 120 Two/Four-Passenger Sport Coupe. The owners won this vehicle in a contest held at the 1935 New York Automobile Show.

Model 120 Two/Four-Passenger
Convertible Coupe.

Packard Eight, Model 1201 Five-Passenger Phaeton.

14

Packard Eight, Model 1201 Two/Four-Passenger Coupe Roadster.

Belgian Prince Eugene de Ligne and 1935 Packard Eight, Model 1201 Five-Passenger Convertible Victoria.

Packard Eight, Model 1201 Five/Six-Passenger Formal Sedan.

Packard Eight, Model 1201 Five-Passenger Club Sedan.

Packard Eight, Model 1201 Five-Passenger Convertible Victoria.

Packard Eight, Model 1201 Two/Four-Passenger Coupe.

Packard Eight, Model 1202 Seven-Passenger Sedan.

Packard Eight, Model 1201 All-Weather Cabriolet by Derham. A converted Five/Seven-Passenger Formal Sedan.

Packard Twelve, Model 1207 Five-Passenger Convertible Victoria.

Packard Twelve, Model 1208 Five-Passenger Convertible Sedan.

Packard Twelve, Model 1208 Seven-Passenger Sedan.

1936 Fourteenth Series

Two new 1936 Packards loaded on an auto transporter. At left, a Model 120-B Five-Passenger Touring Sedan. At right, a Model 120-B Five-Passenger Sedan.

Model 120-B Five-Passenger Convertible Sedan, photographed in Sofia, Bulgaria with the Packard distributor.

Model 120-B Two/Four-Passenger Convertible Coupe.

Model 120-B Two/Four-Passenger Sport Coupe with custom trailer.

Packard Eight, Model 1400 Five-Passenger
Sedan.

Packard Eight, Model 1402 Seven-Passenger Limousine.

Packard Eight, Model 1401 Two/Four-Passenger Coupe Roadster.

Packard Super Eight, Model 1407 Five-Passenger Coupe.

Packard Super Eight, Model 1404 Five-Passenger Club Sedan.

Packard Twelve, Model 1407 Five-Passenger Club Sedan.

Packard Twelve, Model 1407 Five-Passenger Phaeton. In background is a Packard Twelve, Model 1407 Five-Passenger Club Sedan, photographed at the Packard Proving Grounds.

Packard Twelve, Model 1408 Seven-Passenger Limousine.

1937 Fifteenth Series

Packard Six, Model 115-C Five-Passenger Club Sedan.

Packard Six, Model 115-C Five-Passenger Sedan.

Packard Six, Model 115-C Special Panel Delivery.

Packard Six, Model 115-C Two-Passenger Business Coupe.

Model 120-C Five-Passenger Touring Sedan.

Model 120-C Two/Four-Passenger Convertible Coupe. This was a retouched photograph of a 1936 Model 120-B.

Model 120-C Five-Passenger Station Wagon by Brooks Stevens.

Model 120-C Five-Passenger Convertible Sedan.

Packard Super Eight, Model 1500 Five-Passenger Touring Sedan.

Ballet Russe star Yurek Shabelevsky posed with Packard Super Eight, Model 1501 Five-Passenger Convertible Victoria.

Packard Super Eight, Model 1501 Two/Four-Passenger Coupe Roadster.

Packard Super Eight, Model 1501 Two/Four-Passenger Coupe.

Packard Super Eight, Model 1501 Five-Passenger Coupe.

Packard Twelve, Model 1507 Five/Seven-Passenger Formal Sedan.

Packard Twelve, Model 1506 Five-Passenger Touring Sedan.

Packard Twelve, Model 1507 Five-Passenger Club Sedan.

Packard Twelve, Model 1507 Five-Passenger Coupe.

Packard Twelve, Model 1507 Five-Passenger Convertible Victoria.

Packard Twelve, Model 1508 Five-Passenger Convertible Sedan.

Packard Twelve, Model 1507 Five-Passenger Club Sedan.

1938 Sixteenth Series

1938 Packard models at the entrance to the proving grounds. At left, a Packard Eight, Model 1601 Five-Passenger Touring Sedan. At right, a Packard Six, Model 1600 Five-Passenger Touring Sedan.

Packard Six, Model 1600 Five-Passenger Touring Sedan. The gentleman in the center won the vehicle in a contest. He took delivery from the Toronto Packard dealership.

Prototype Packard Six, Model 1600 Five-Passenger Touring Sedan. This vehicle undergoing weight strength test at the Utica proving grounds.

Packard Six, Model 1600 Two/Four-
Passenger Club Coupe.

Packard Eight, Model 1601
Two-Passenger Business Coupe.

Interior of a Packard Eight, Model 1601-D 5-Passenger DeLuxe Touring Sedan. Packard approved accessories included: DeLuxe steering wheel; DeLuxe gear shift knob; cigar lighter; clock.

Packard Eight, Model 1601-D Five-Passenger DeLuxe Touring Sedan.

Interior detail of a Packard Eight, Model 1601 Five-Passenger Touring Sedan.

Packard Eight, Model 1601 Five-Passenger Touring Sedan.

Packard Eight, Model 1601 Five-Passenger Touring Sedan and Burd Piston Ring Special racer. Driven by Floyd Roberts and powered by a 270 cubic inch Miller 4-cylinder engine, the Burd Piston Ring Special finished first at the 1938 Indianapolis 500.

Experimental Packard Eight, Model 1601 Two/Four-Passenger Club Coupe built for Mr. Edward Macauley. The radiator louvers were adopted on the 1939 Super Eight and Twelve; front fenders on the 1939 120 and Super Eight.

Packard Eight, Model 1601 Five-Passenger 2-Door Touring Sedan.

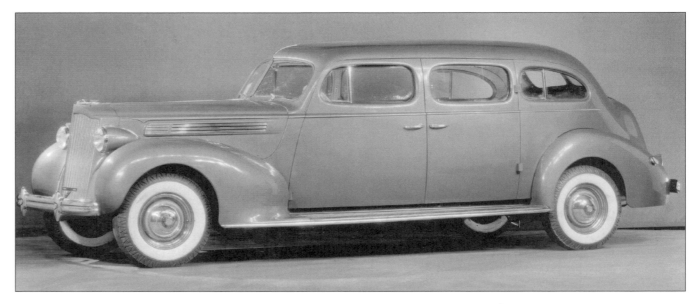

Packard Eight, Model 1602 Five/Seven-Passenger Touring Sedan.

Packard Eight, Model 1601 Two/Four-Passenger Club Coupe.

Packard Eight, Model 1601 Two/Four-Passenger Convertible Coupe.

Packard Eight, Model 1601 Five-Passenger Convertible Sedan.

Packard Eight, Model 1601 Seven-Passenger All-Weather Panel Brougham by Rollston.

Packard Super Eight, Model 1603 Five-Passenger Touring Sedan and Mr. Alvan Macauley in front of his home, Grosse Pointe, Michigan.

Packard Super Eight, Model 1604 Five-Passenger Club Sedan.

Packard Super Eight, Model 1605 Five/Seven-Passenger All-Weather Town Car by Rollston.

Packard Twelve, Model 1607 Five-Passenger Touring Sedan.

Packard Twelve, Model 1607 Five/Seven-Passenger All-Weather Cabriolet by Rollston.

Packard Twelve, Model 1608 Five/Seven-Passenger All-Weather Cabriolet by Brunn.

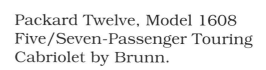

Packard Twelve, Model 1608
Five/Seven-Passenger Touring
Cabriolet by Brunn.

1939 Seventeenth Series

Packard Motor Car Company 40th anniversary display. Foreground, Packard Six, Model 1700 Two/Four-Passenger Club Coupe. Center left, Super Eight, Model 1703 Two/Four-Passenger Convertible Coupe. Top right, Packard 120, Model 1701 Five-Passenger Touring Sedan.

Packard Six, Model 1700 Eight-Passenger Station Wagon by Cantrell.

Packard Six, Model 1700 Five-Passenger Touring Sedan.

Left, Packard Six, Model 1700 Five-Passenger Touring Sedan. Right, Packard 110, Model 1800 (1940) Five-Passenger Touring Sedan.

Packard 120, Model 1701 Two/Four-Passenger Club Coupe, photographed in front of The Greenbrier, White Sulpher Springs, West Virginia.

Packard 120, Model 1701 Two/Four-Passenger Convertible Coupe.

Packard 120, Model 1701 Two/Four-Passenger Convertible Coupe.

Packard Super Eight, Model 1703 Five-Passenger Touring Sedan.

Packard Super Eight, Model 1703 Two/Four-Passenger Convertible Coupe. This car was modified with three-way Coupe de Ville top and landau irons, and leather and whipcord upholstery for Mr. Edward Macauley.

Packard Super Eight, Model 1703 Five-Passenger Convertible Sedan.

Packard Twelve, Model 1708 Five/Six-Passenger Parade Car by Rollson. Passengers included US President Franklin Delano Roosevelt and British King George VI, in rear seat, and Sir Ronald Lindsay, British Ambassador to the United States, in jump seat.

Packard Twelve, Model 1708 Five/Seven-Passenger All-Weather Cabriolet by Brunn.

Packard Twelve, Model 1707 Five/Seven-Passenger All-Weather Cabriolet by Rollson.

Packard Super Eight, Model 1703 Convertible Victoria by Darrin.

1940 Eighteenth Series

Packard 120, Model 1801 Five-Passenger Touring Sedan, photographed at the Packard Photographic Laboratory.

Packard 110, Model 1800 Two/Four-Passenger Club Coupe.

Packard 110, Model 1800 Two/Four-Passenger Convertible Coupe.

Packard 110, Model 1800 Two-Passenger Business Coupe.

Prototype Packard 120, Model 1801 Five-Passenger Convertible Sedan.

Packard 120, Model 1801 Five-Passenger Touring Sedan.

Packard 120, Model 1801 Two/Four-Passenger Club Coupe.

Packard 120, Model 1801 Five-Passenger Touring Sedan.

Packard 120, Model 1801 Five-Passenger Club Sedan.

Packard Super Eight 160, Model 1803 Five-Passenger Club Sedan.

Packard Super Eight 160, Model 1803 Five-Passenger Convertible Coupe.

Packard Super Eight 160, Model 1803 Five-Passenger Convertible Sedan.

Packard Custom Super Eight 180, Model 1806 Convertible Victoria by Darrin, photographed at the 40th annual New York Automobile Show.

Packard Custom Super Eight 180, Model 1806 Convertible Victoria by Darrin.

1941 Nineteenth Series
and
1942 Twentieth Series

Two views of a 1941 Packard 110, Model 1900 Eight-Passenger Station Wagon by Hercules Body Company.

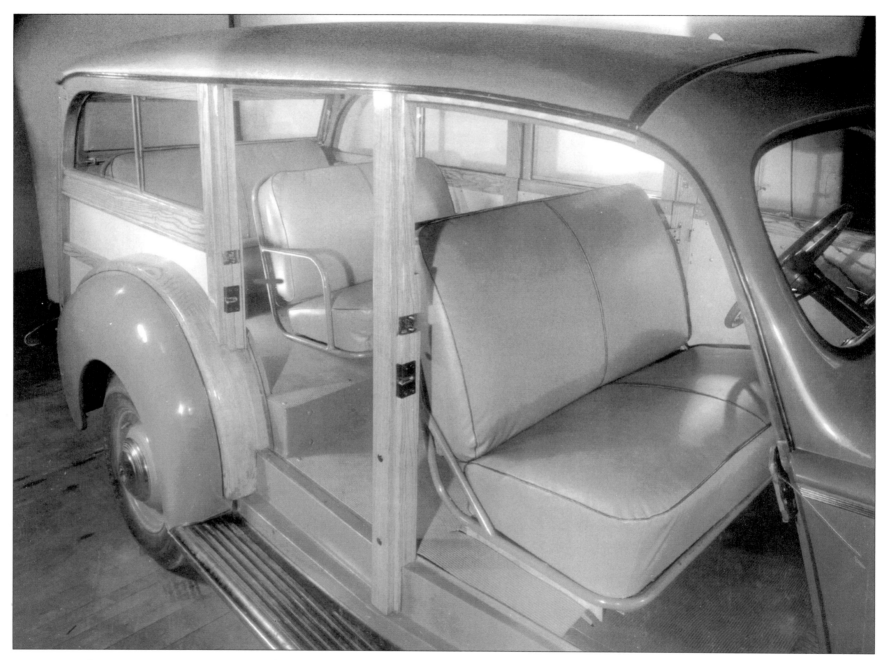

Interior of a Hercules Body Company Eight-Passenger Station Wagon.

1941 Packard 110, Model 1900 Two/Four-Passenger Convertible Coupe.

Pre-production 1941 Packard 110, Model 1900 Five-Passenger Convertible Coupe.

1941 Packard Super Eight Custom 180, Model 1908 Limousine by LeBaron.

An early version of Edward Macauley's Phantom, with conversion executed in the Packard shops from a 1941 Packard Super Eight Custom 180 Convertible Victoria.

1942 Packard Super Eight Custom 180, Model 1908 Limousine by LeBaron. Photo appeared in the Twentieth Series catalogue.

Two views of a prototype of the 1942 Packard 110, Model 2020 Five-Passenger Convertible Coupe.

1942 Packard 120, Model 2021 Five-Passenger Custom Convertible Coupe.

Two 1942 Packard Custom Super Eight 180, Model 2008 Seven-Passenger Sedans by LeBaron. The last 1942 Packards sold to the Russian government.

The Iconografix Photo Archive Series includes:

AUTOMOTIVE

AMERICAN SERVICE STATIONS 1935-1943 *Photo Archive*	ISBN 1-882256-27-1
IMPERIAL 1955-1963 Photo Archive	ISBN 1-882256-22-0
IMPERIAL 1964-1968 Photo Archive	ISBN 1-882256-23-9
LE MANS 1950: THE BRIGGS CUNNINGHAM CAMPAIGN *Photo Archive*	ISBN 1-882256-21-2
PACKARD MOTOR CARS 1935-1942 Photo Archive	ISBN 1-882256-44-1
PACKARD MOTOR CARS 1946-1958 Photo Archive	ISBN 1-882256-45-X
PHILLIPS 66 1945-1953 Photo Archive	ISBN 1-882256-42-5
SEBRING 12-HOUR RACE 1970 Photo Archive	ISBN 1-882256-20-4
STUDEBAKER 1933-1942 Photo Archive	ISBN 1-882256-24-7
STUDEBAKER 1946-1958 Photo Archive	ISBN 1-882256-25-5

TRUCKS

DODGE TRUCKS 1929-1947 Photo Archive	ISBN 1-882256-36-0
DODGE TRUCKS 1948-1960 Photo Archive	ISBN 1-882256-37-9
STUDEBAKER TRUCKS 1927-1940 Photo Archive	ISBN 1-882256-40-9
STUDEBAKER TRUCKS 1941-1964 Photo Archive	ISBN 1-882256-41-7

AVAILABLE EARLY 1996

COCA-COLA: A HISTORY IN PHOTOGRAPHS 1930-1969	ISBN 1-882256-46-8
COCA-COLA: ITS VEHICLES IN PHOTOGRAPHS 1930-1969	ISBN 1-882256-00-X

The Iconografix Photo Archive Series is available from direct mail specialty book dealers and bookstores worldwide, or can be ordered from the publisher. For additional information or to add your name to our mailing list contact:

Iconografix
PO Box 609
Osceola, Wisconsin 54020 USA

Telephone: (715) 294-2792
(800) 289-3504
Fax: (715) 294-3414

Book trade distribution by Voyageur Press, Inc., PO Box 338, Stillwater, Minnesota 55082 USA   (800) 888-9653

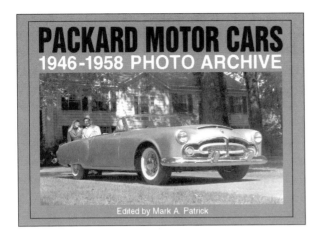

PACKARD MOTOR CARS
1946-1958 PHOTO ARCHIVE
Edited by Mark A. Patrick

MORE
GREAT BOOKS FROM
ICONOGRAFIX

PACKARD MOTOR CARS 1946-1958
Photo Archive ISBN 1-882256-45-X

AMERICAN SERVICE STATIONS 1935-1943
Photo Archive ISBN 1-882256-27-1

**LE MANS 1950: THE BRIGGS CUNNINGHAM
CAMPAIGN** *Photo Archive* ISBN 1-882256-21-2

DODGE TRUCKS 1929-1947 *Photo Archive*
ISBN 1-882256-36-0

IMPERIAL 1955-1963 *Photo Archive*
ISBN 1-882256-22-0

**COCA-COLA: ITS VEHICLES IN PHOTO-
GRAPHS 1930-1969** ISBN 1-882256-00-X

STUDEBAKER 1933-1942 *Photo Archive*
ISBN 1-882256-24-7

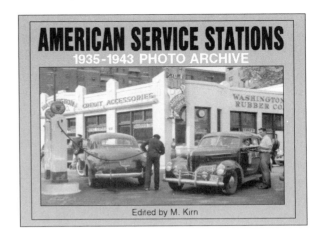

AMERICAN SERVICE STATIONS
1935-1943 PHOTO ARCHIVE
Edited by M. Kirn

LE MANS 1950 PHOTO ARCHIVE
The Briggs Cunningham Campaign
Edited with introduction by Robert C. Auten • Photographs by Smith Hempstone Oliver

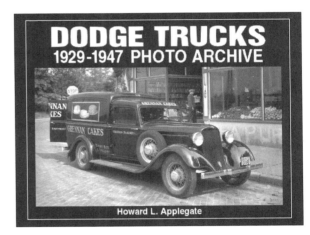

DODGE TRUCKS
1929-1947 PHOTO ARCHIVE
Howard L. Applegate

IMPERIAL 1955-1963
PHOTO ARCHIVE
Edited by P. A. Letourneau

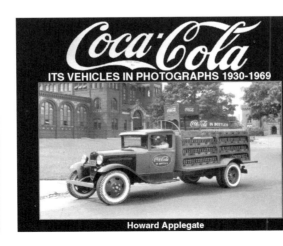

Coca-Cola
ITS VEHICLES IN PHOTOGRAPHS 1930-1969
Howard Applegate

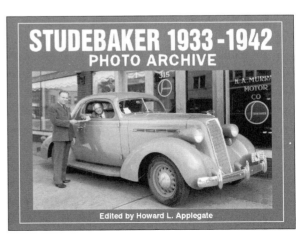

STUDEBAKER 1933-1942
PHOTO ARCHIVE
Edited by Howard L. Applegate